GROWING PAINS!

HOW I GREW HEMP

DORA WILSON-JEFFERSON

ISBN: 978-1-7366366-0-2 (E-Book)

ISBN: 978-1-7366366-1-9 (Paperback)

GROWING PAINS!

Printed in the United States of America

CONTENTS

INTRODUCTION

G rowing Hemp has been an adventure and a challenge. There are many different preparations and methods that must be considered when using virgin land (land that is not cultivated) versus farmland (land that has had crops planted there in previous years).

What are you growing your crop for?

The reason I decided to grow my crop is to provide a farm commodity for manufacturers' and consumers' health. First, learn all you can about the product, which requires extensive and tedious research. Next, create a plan for how to carry out your production, and then, find a buyer to sell your product to. Furthermore, make sure you **have a signed agreement from the purchaser before you plant.**

Hemp just recently became legally grown in our region and it is used for many different purposes. Know what product your hemp will be used in. Hemp is an ingredient and CBD is the oil produced by the hemp plant. This oil can be used in many products due to its many properties derived from different varieties of plants. There is the Oil, the Smokable flower, and the Biomass. Each is used in a different capacity for health benefits.

Here are a few products that CBD is being used in as an ingredient:

1. Pain creams
2. Oils (sleep-anxiety-pain oils)
3. Human and animal clothing for calming effect
4. Cigarettes (snuff-dip-tobacco patches)
5. Pet food

CHAPTER ONE

PLANTING—WHAT YOU WILL NEED TO GET THE PROJECT STARTED

To keep things short, I will get straight to the point. There is a process to prepare yourself to enter the world of farming hemp. First there are things you should do to prepare yourself before considering this venue. My motto for any project that involves money, or investment of time and energy, reflects on a scripture that says; "before a man builds a house he must count the cost (Luke 14:28)." So before I began any type of funding for this project, I did some research.

This research entailed attending many farm meetings to get information that was available to us as farmers prior to us making a decision as to whether we would be up to the task of putting this project into motion. There were delays for obtaining the permit for the state of Georgia which caused us to have a late fall crop. So, we were given our application in June 2020 and we began the process in July of 2020.

There are three things you must have before starting a hemp-growing project...

1. **Time**: Be prepared to spend a lot of time in the field. Monitor your health well. Hire others to assist you as needed.

2. **Land**: YYou must have a deed to the property or a lease from the person who holds the deed of record to the area that you plan to plant.

3. **Money**: Be prepared to structure a plan for upfront costs...expenditures during the grow process and the cost to harvest. Your processor will deduct their fees from your product after it has been processed.

If there is a deficit of either, then you run a risk of not fulfilling your obligation in bringing your crop to a full, viable state. **Make sure you plan!** If you fail to plan, you plan to fail. It might be wise to have a backup plan and in certain instances, a backup to your backup plan.

To plant a viable crop; you must do your research and **"Count the Cost."** Attend as many farm meetings as possible or webinars concerning hemp that are available in your farming community—preferably via zoom or online. Subscribe to available informational websites concerning the product; doing so will provide you with current and up to date information. I **RECOMMEND MY BOOK!!!!**

1. Make sure that you have a farm number if you are new to farming.

2. If you plan to purchase farmland, then check the zoning laws in your city. Property must be zoned "agricultural." I found this out the hard way by making a purchase of land only to find out that the tract needed to be at least 40 acres to be rezoned as "agricultural." *DO YOUR RESEARCH!!!*

3. To be a farmer requires a certain amount of tenacity and a purpose. Hard work and long hours are not attractive to the naked eye, but I set long-term goals for myself and my family, you should as well. My choice to venture into these uncharted waters as a grower of hemp was based on a life-long desire of mine. This is to leave a legacy for my immediate family and generations to come. There are some things in this world that will remain as long as the earth remains. Pain is one of those things! A tried-and-true resolution to many problems with pain is the properties found in the hemp plant. *I believe Hemp will stand the test of time!*

This is a guide based on experience and knowledge to assist any person who considers growing hemp in the climate challenges I endured in Georgia. Current farmers, as well as future farmers, may find this information helpful in deciding whether to take the quantum leap into the new age venture of Hemp Farming. Here is where I share the challenges of the process that I experienced and provide a guide to plant, grow, and harvest the product.

1. Estimated cost per acre to grow: $25,000-$40,000.

2. Area to grow first time: 5-10 acres.

3. Soil samples, for growers of tomatoes, will require the same level of nutrients to grow hemp.

4. Locate service providers to install irrigation-system well, if needed, and lay plastic for weed control.

5. Hemp planting will require 1200-2000 plants per acre.

6. Know what a male plant looks like.

7. Purchase clones that are 99% defeminized.

8. Have enough manpower to walk rows daily to check irrigation, weed growth, and insect control.

9. Migrant workers can be hired, but make sure they know your expectations.

10. Comply with protocol mandated by the state to have samples tested around 10 weeks to maintain CBD and THC level (panel test).

Things you need:

Obtaining the hemp License from the state is a process that will take approximately 30 days or more. This is based on the time of year that you apply. The licensing department will get an overflow of applications in January and February due to renewals and late filers. The license will expire on December 31 of each year. The requirements to be eligible for the Hemp License is to pass a background check, submit fingerprints and show proof of control of the farm area that you intend to grow on. This can be verified with a deed or lease agreement to the property.

1. You must have a farm number to apply for a hemp license—GA Department of Agriculture.

2. You must have documentation from the landowner (lease or deed).

3. You must provide grid coordinates of farm location.

4. A Criminal background check must be obtained.

5. The nutrients needed for your plant varieties, frequencies, and amounts must be coordinated.

6. Water Pressures from well to field should be checked (pressure valves).

7. The best Preparation for aggressive weed control methods is pulling them by hand.

8. There will be many insects that will challenge your crop (ants, roaches, caterpillars, etc.)

9. Locate and contract a processing facility that you handle your harvest and purchase your crop.

10. Keeping your buyer in the loop as you get closer to harvest is essential.

11. Always have a signed agreement from purchaser before you plant.

Researching the type of plant variety, you plan to grow in your region is going to be key. If you are growing for CBD, which we all should be, you want to select a variety that yields a high level of CBD and is resistant to the weather. Keeping in mind it will take you from 90 to 120 days (3-4 months) for the total process. Therefore, when you plant your crop is important so that your harvest time is during a time when the weather will be sufficient for the plants to grow and receive the highest yield. Based on my research and experience you can expect to lose about 30% of the plants you purchase. For every 2,000 plants you purchase it would be expected to lose at least 600 of those plants.

You must have your soil sampled to check the level of nutrients, based on the crop that will be planting. Your county extension agency can provide soil-sample test kits that will be forwarded to your local agricultural university for testing. They will return your sample results in a timely manner, so you can approach your local farm co-op for a cost analysis of the nutrients that are needed to bring soil up to the desired standards.

You will need to provide the number of acres that you will be farming to get proper calculations. I suggest that any farmer plant 3-10 acres for the first time when farming hemp. This will allow you to get a feel of the time and attention required to get your plants to harvest. These plants require a lot of attention!

Chapter Two

Growing—My adventures and challenges growing Hemp for the first time

The greatest challenge in the hemp growing process was manpower. I called this a challenge because I am a 60-year-old woman who retired from the medical profession. As a nurse, my normal thought process was to nurture and care for the lives of those that were placed in my care. When the transition was required to care for a growing plant that could not speak, I was at a disadvantage in trying to determine the needs of the plant and provide what was required of me to keep it viable until it could be harvested. Every living organism requires nutrition and must be maintained to prevent its demise. Your Hemp must receive proper nutrients or you will lose your crop.

A solution going forward would be to include manpower in the early planning stage. I recommend calculating 15-20 hours per week per acre to determine the manpower that will be needed to maintain your crop. This labor can range from 13- 20 dollars per hour, which is customary. During harvest week you can calculate at 70 hours of manpower per acre based on 8-hour days. For example, 2.5 persons per week would be needed to maintain your operation on a 5-acre tract during the 12-15 week grow process.

In the life of the hemp plant there are many insects that love the taste, smell, and texture of the plant, for this reason you will find the insects to be a nuisance to the farmer. They must be eradicated, or the farmer will experience the pain of losing their investment. The greatest pain may be experienced through the loss of the crop that they have planted with the intent to enjoy a profit. Pain not only can be felt from the infestation of insects it can be felt when weeds that receive the same nutrients that are intended for the hemp plants overtake your growing area. It can be especially challenging to regain control of your field if weeds or insects overtake your crop. I found that growing in South Georgia's, humid weather and unexpected rain can contribute to insect infestation. Manpower is of the essence! This is not a feat that can be achieved with tractors or plows.

Once you have laid down the plastic covering to help control the weeds and hold moisture in your soil, the weeds will also enjoy the same benefits of the nutrients that you have provided for your plants. Many farmers have become creative with ideas of how to control the weed overgrow. By using mechanical equipment such as weed eaters or tillers and even plows, I found this to be very exhausting. If you decide to use a zero turn to manage weeds, which is conducive when you use the proper spacing of your rows. Be sure to keep your mower guard in place to allow debris to fall to the ground. Otherwise, debris blowing on your plants can cause the plants to become diseased. The hemp plant is such a delicate plant requiring tedious care and attention to keep the buds intact as your plants grow. I found that this idea may produce a large amount of biomass but as a farmer, you must be aware of what will bring in the best return on your investment.

For those interested in the indoor growing of hemp there are some things to keep in mind. Hemp is a plant that requires you to control the CBD levels. Heat brings out the CBD and THC levels. When utilizing a greenhouse, with a controlled temperature, your plant may mature quicker and have a higher level before the plants reach maturity. South Georgia's indoor growing facilities have not proven to be the best route.

My experience is with an outdoor grow and this is the basis for my story. Thus, knowing the length of time for the variety of plant you have selected is just as important as the other variables. As it would be detrimental to grow a plant that matures in 8 weeks and it stays in the field for 12 weeks. Why? You have set yourself up for the probability of having a THC level higher than what is accepted by the Department of Agriculture. This may cause you to lose your entire crop because regulation requires the crop to be burned when the THC level exceeds .3. There is a variable that can be calculated that will give leeway to that number. You must consult with your Department of Agriculture.

As a hemp grower your main objective is to grow your crop to be used in a product that will pay the farmer the best return on his investment. Therefore, another rule of thumb is to not invest more than you are willing to lose. I have researched some of the products that are currently on the market and according to my research "smokable flower" is the most lucrative product that a farmer can process to be sold to the processor or manufacturer. Always have your product sold before you plant!

CHAPTER THREE

THE HARVEST—NOW THAT YOU HAVE COMPLETED YOUR 12 TO 15 WEEKS OF GROWING HEMP, WHAT IS NEXT?

After the state auditors have cleared your crop with a THC level that is within the state regulations, you will have 15 days according to the state of Georgia Department of Agriculture to harvest your crop. What does it mean to harvest? Harvesting is to remove your plants from the soil. The process can be accomplished by either cutting them down or pulling them up by the roots.

There is a difference in harvesting by pulling or cutting and knowing what you are growing your plants to produce will determine which method to use during harvest. If you are growing to produce biomass or oil you can root pull your plant. However, if you are growing to produce smokable flowers it is better to cut your crop with a machete or cane cutter to preserve the integrity of your flowers. Even though my direct participation during this time was limited, there was a major challenge dealing with the unpredictable weather in Georgia. The weather contributed to having to divert from the normal process and my creativity was a powerful tool to possess during that time.

Once the plant is free from the soil you will have two options to move forward with your process. The first option will be to have the plants transferred to a pre-assigned processing plant for further processing. The second option is to obtain a processor's license and complete the processing in your own facility. The processing of hemp has many components and is not a task, based on my experience that a first-time hemp farmer should take on during their first hemp grow out. So, identifying and selecting a processor is crucial. If you do not know what your processor is equipped to do with your product you may have a disaster. If your processor can only process oil or biomass and you have grown the variety of plant conducive for smokable flowers you will be disappointed with your payout. Knowing what your processing facility is equipped to do is a major component to planning when growing hemp.

As a farmer it is also important to know something about transportation and the process of getting your product from the field to the processor. By law we are legally required to have a manifest to transport from point A to point B. An alternative to transporting promptly is to store your product in a refrigerated trailer until it can be taken to the processor. It is also vital to know your dry weight and wet weight because this calculation is what your pay rate will be based on. Your wet weight is the weight of the product after harvest. Dry weight may take several weeks after reaching the processor which gives you the number of pounds for which you will be compensated. There is a different pay scale for Hemp oil, biomass and smokable flower.

In writing this material, I want to acknowledge that the success I had on this journey was not an independent effort. I had help from various sources that assisted and provided me with the necessary resources to obtain the maximum return. As previously mentioned, doing your research is key in every stage of the project. Especially, when it comes to identifying your own distributor or marketing firm. This is to ensure that you select a credible company that has reliable references. This can be a challenge even with doing your research. This is why it is strongly suggested that you take the time to at least consult with an attorney if you don't have one. Having an attorney to review documents before signing can be vital in the long run.

This is my story about the challenges and why I decided to take on this project. It is my wish that you will decide to venture into your own hemp-growing adventure and that it be a prosperous and lucrative one. This is my "WHY," but everyone must consider their own WHY for this project.

18

My name is Dora Wilson Jefferson formerly Dora Ann Jackson of Hawkinsville, Georgia where I received my formal education. I'm currently residing in Albany, Ga where I have lived for 36 years. I am a mother to 6 adult children, 4 step children, 7 grandchildren, and 1 great grandchild. My life, as I remember it, began living the farm life with my mother and 6 siblings, in rural Pulaski County, Ga. We were a part of the community of workers in the south, who worked in the fields. I recall as a preschooler having to be in the field with my mother because there were no daycares or child care services during those times.

From that point I learned to pick cotton and be productive. My mother fashioned a small cotton sack from a flour sack for me to use. I worked alongside her, picking cotton, until I was five and started to school. Life had always been a struggle, but by the grace of God, I was able to grow and prosper. I married my first husband in Hawkinsville, Georgia and was blessed with 6 beautiful children. Later, I moved to Albany, Georgia and continued to work many different jobs, including in the medical field.

Poverty made hard work a forte for me because my mindset saw it as a way to survive as well as rise up. As an adult I created a level playing field for myself and my children by working multiple jobs. This was to provide a better life for them than the one I had. It gave me the drive to do better, and I wanted to instill this in my children. I wanted to provide them with the things I was not able to enjoy.

My story is one of bitter-sweet experiences with a miraculous outcome. The many hurdles and hardships I have endured taught me that there is always a lesson to be learned. This is the most valuable lesson I have remembered and learned from all of those challenges. So, as I ventured into adulthood I did have an encounter with the government that caused me to spend a short, but well needed rest from society for a period of 3 years. Even under those circumstances, my mindset was to advance and I was shown favor with the prison guards and warden, who allowed me to excel and prosper while in the enemy's camp.

Throughout this time, I developed a greater love for my freedom as well as an appreciation for my family, mainly my mother who instilled in me self-worth and tenacity. This was a time when the spiritual gift that she transferred to me when she passed became relevant and real. I was endowed the courage and strength to carry on when others would have given up. My Loved ones were missing in my environment, but always present in my heart. "Love is the greatest gift of all." I attribute my strength to a song written by Whitney Houston during this period of separation.

Fast forwarding to the present, my struggles and successes have been many. I was introduced to a young man who is a retired Vietnam Veteran, whom I married and he gave me his last name of Jefferson. He gave me an opportunity to use my God-given gift of a supernatural desire to achieve all that was intended for me in this life through grace and a perfect way. I thank God for my children, my family, my husband and all those people who rallied around me in this effort to share my challenges and success.

In spite of the Covid-19 pandemic, the sluggish economy, and the troubled government system, my being able to grow hemp in southwest Georgia in the year 2020 is another monumental blessing. This was a time for trial and error, since 2020 is the first year hemp has been allowed to be grown in Georgia. Therefore, I share in this booklet the basic requirements for anyone who decides they would like to grow hemp in Georgia.

In life I've learned that with persistence and ambition you will make it to your destination! I encourage you to plant your seed in fertile ground and stay the course to reap a bountiful harvest. "For ambition is the path to success and persistence is the vehicle you arrive in." –Bill Bradley

"Knowing is not enough you must apply, being willing is not enough you must do." –Johann Wolf Gang

ACKNOWLEDGEMENTS

I would like to acknowledge the Southwest Georgia Project of Albany GA. Especially, Ms. Karen Lawrence, my Ag specialist who guided me through many challenges over the years. How would I have made it without manpower? Therefore, I acknowledge the team of migrant workers and day laborers for their hard work and dedication to the project. As well as anyone that consulted with us and assisted in monitoring our progress. You all were a true blessing. I am grateful to you all. WE DID IT!

Hemp Plant

Male Plant vs Female Plant

Female Hemp Plant

Male Hemp Plant

Insect Infested Hemp Plant

Dry Rotted Hemp Plant

HEMP LABORATORY TEST

CERTIFICATE OF ANALYSIS

SC Laboratories, LLC
100 Pioneer Street, Suite E
Santa Cruz, CA 95060
(866) 435-0709 | sclabs.com

Sample Name:	CBG		Date Collected:	09/30/2019
LIMS Sample ID:	190930T022		Date Received:	09/30/2019
Batch #:	1		Tested for:	KB Hemp Co
Sample Metrc ID:			License #:	
Sample Type:	Flower, Inhalable		Address:	
Batch Count:			Produced by:	
Sample Count:			License #:	
Unit Mass:			Address:	
Serving Mass:				
Density:			Overall result for batch 1: Pass	

Moisture Test Results   10/01/2019

	Results (%)
Moisture	10.7

Water Activity Test Results

	Results (Aw)	Action Limit Aw

Cannabinoid Test Results   10/01/2019

Cannabinoid analysis utilizing High Performance Liquid Chromatography (HPLC, QSP 5-4-4-4)
Calculated using Dry-Weight

	mg/g	%	LOD / LOQ mg/g
Δ9THC	ND	ND	0.052 / 0.158
Δ8THC	ND	ND	0.074 / 0.224
THCa	1.690	0.1690	0.052 / 0.156
THCV	ND	ND	0.045 / 0.137
THCVa	ND	ND	0.088 / 0.267
CBD	ND	ND	0.059 / 0.180
CBDa	0.254	0.0254	0.052 / 0.156
CBDV	ND	ND	0.027 / 0.080
CBDVa	ND	ND	0.030 / 0.090
CBG	2.075	0.2075	0.048 / 0.144
CBGa	183.096	18.3096	0.034 / 0.102
CBL	ND	ND	0.114 / 0.346
CBN	ND	ND	0.052 / 0.157
CBC	0.170	0.0170	0.048 / 0.146
CBCa	4.021	0.4021	0.233 / 0.705
Sum of Cannabinoids:	**191.306**	**19.1306**	
Total THC (Δ9THC+0.877*THCa)	1.482	0.1482	
Total CBD (CBD+0.877*CBDa)	0.223	0.0223	

	Action Limit mg
Δ9THC per Unit	
Δ9THC per Serving	

Batch Photo

Terpene Test Results   10/02/2019

Terpene analysis utilizing Gas Chromatography - Flame Ionization Detection (GC - FID)

	mg/g	%	LOD / LOQ mg/g
α Pinene	<LOQ	<LOQ	0.028 / 0.084
Camphene	ND	ND	0.038 / 0.116
Sabinene	ND	ND	0.024 / 0.073
β Pinene	ND	ND	0.016 / 0.048
Myrcene	0.853	0.0853	0.03 / 0.092
α Phellandrene	ND	ND	0.048 / 0.144
3 Carene	ND	ND	0.028 / 0.085
α Terpinene	ND	ND	0.051 / 0.155
Limonene	0.16	0.016	0.04 / 0.12
Eucalyptol	ND	ND	0.051 / 0.155
Ocimene	ND	ND	0.053 / 0.16
γ Terpinene	ND	ND	0.038 / 0.114
Sabinene Hydrate	ND	ND	0.046 / 0.138
Fenchone	ND	ND	0.06 / 0.181
Terpinolene	ND	ND	0.042 / 0.128
Linalool	ND	ND	0.043 / 0.13
Fenchol	ND	ND	0.051 / 0.153
(-)-Isopulegol	ND	ND	0.026 / 0.08
Camphor	ND	ND	0.08 / 0.242
Isoborneol	ND	ND	0.028 / 0.085
Borneol	ND	ND	0.063 / 0.19
Menthol	ND	ND	0.043 / 0.129
Terpineol	ND	ND	0.029 / 0.087
Nerol	ND	ND	0.042 / 0.128
R-(+)-Pulegone	ND	ND	0.016 / 0.047
Geraniol	ND	ND	0.037 / 0.112
Geranyl Acetate	ND	ND	0.025 / 0.076
α Cedrene	ND	ND	0.012 / 0.035
β Caryophyllene	0.339	0.0339	0.029 / 0.087
α Humulene	0.076	0.0076	0.017 / 0.051
Valencene	<LOQ	<LOQ	0.018 / 0.055
Nerolidol	<LOQ	<LOQ	0.05 / 0.15
Caryophyllene Oxide	ND	ND	0.011 / 0.034
Guaiol	0.770	0.0770	0.035 / 0.106
Cedrol	ND	ND	0.022 / 0.066
α Bisabolol	0.507	0.0507	0.057 / 0.172
Total Terpene Concentration:	**2.705**	**0.2705**	

Sample Certification

California Code of Regulations Title 16 Effect Date January 16, 2019
Authority: Section 26013, Business and Professions Code
Reference: Sections 26100, 26104

Josh Antunovich, LQC Verified By
Date: 10/02/2019

Josh Wurzer, President
Date: 10/02/2019

Hemp Lab Test Certificate

NOTES

NOTES

NOTES

NOTES

NOTES

NOTES